Particle and Wave

BENJAMIN LANDRY

Particle and Wave

THE UNIVERSITY OF CHICAGO PRESS

Chicago & London

BENJAMIN LANDRY is a Meijer Post-MFA Fellow at the
University of Michigan.

The University of Chicago Press, Chicago 60637
The University of Chicago Press, Ltd., London
© 2014 by The University of Chicago
All rights reserved. Published 2014.
Printed in the United States of America
23 22 21 20 19 18 17 16 15 14   1 2 3 4 5

ISBN-13: 978-0-226-09619-3 (paper)
ISBN-13: 978-0-226-09622-3 (e-book)
DOI: 10.7208/chicago/9780226096223.001.0001

Library of Congress Cataloging-in-Publication Data

Landry, Benjamin, author.
    [Poems. Selections]
    Particle and wave / Benjamin Landry.
        pages ; cm. — (Phoenix poets)
    Poems.
    ISBN 978-0-226-09619-3 (pbk. : alk. paper) —
    ISBN 978-0-226-09622-3 (e-book)
    I. Title. II. Series: Phoenix poets.
PS3612.A5487A6 2014
811'.6—dc23

                                    2013021972

⊚ This paper meets the requirements of ANSI/NISO Z39.48-1992
(Permanence of Paper).

For my parents,
John and Adrienne

CONTENTS

Acknowledgments ix

H	[Hydrogen]	*3*
He	[Helium]	*4*
Li	[Lithium]	*5*
C	[Carbon]	*6*
N	[Nitrogen]	*8*
O	[Oxygen]	*10*
F	[Fluorine]	*12*
Na	[Sodium]	*13*
Mg	[Magnesium]	*15*
Al	[Aluminum]	*17*
Si	[Silicon]	*19*
P	[Phosphorus]	*21*
K	[Potassium]	*22*
Ti	[Titanium]	*23*
Cr	[Chromium]	*24*
Co	[Cobalt]	*26*
Ni	[Nickel]	*28*
Ga	[Gallium]	*29*
As	[Arsenic]	*30*
Br	[Bromine]	*32*
Rb	[Rubidium]	*34*
Sr	[Strontium]	*35*
Tc	[Technetium]	*37*
Sn	[Tin]	*38*
I	[Iodine]	*39*

Ba	[Barium]	*40*
Ce	[Cerium]	*42*
Tm	[Thulium]	*44*
Hf	[Hafnium]	*46*
Au	[Gold]	*48*
Bi	[Bismuth]	*49*
Po	[Polonium]	*50*
At	[Astatine]	*52*
U	[Uranium]	*54*
Cf	[Californium]	*56*
Lr	[Lawrencium]	*57*
Db	[Dubnium]	*59*
Sg	[Seaborgium]	*60*
Hs	[Hassium]	*61*
Uuo	[Ununoctium]	*63*

ACKNOWLEDGMENTS

Thank you to the editors of the following publications for including poems from *Particle and Wave* in their pages:

American Letters and Commentary: "Ga"
Denver Quarterly: "U"
Drunken Boat: "As," "Hf," and "Rb"
Forklift, Ohio: "Lr" and "Tm"
[PANK]: "Uuo"
Sonora Review: "Br" (earlier version)
Web Conjunctions: "Mg," "Po," "Sn," and "Tc"

I would also like to acknowledge the support of a prize granted to an early manuscript of *Particle and Wave* in the Avery Hopwood and Jule Hopwood Contest for 2012 at the University of Michigan.

Special thanks to my English and creative writing teachers, going all the way back to Marlborough Elementary. More recently, thank you to the writing faculty at the University of Michigan who provided valuable critical distance for the poems in this collection: Lorna Goodison, Linda Gregerson, Laura Kasischke, and Keith Taylor. Thank you to the University of Chicago Press production team, to my Michigan cohort, to my friends in writing—Catherine Calabro, Elizabeth Gramm, Kristie Kachler, and Alison Moncrief—and to the principal believers, Sara and Iris.

Particle and Wave

Imagine the heat generated
by Daphne transformed into laurel
and you can begin to feel
what the electron feels
in renouncing its steady orbit.

The new body is a space
humming like a sacristy
or an autoclave,
its language both particle
and wavelength.

Daphne was, of course, an ordinary girl:
desires not especially volatile.
She, too, forgot her terror, nodded off
in the glow of a star appearing
to explode for billions of years.

The rest is medical history,
the 'H' off the interstate ramp.
How the wound of her new life healed.
How the heatshimmer shook
her vision afterward.

2
He
4.0026

Setter of things into motion.

Fractional, what is your subsequent charm?

We divine your nature by experiment:

                    Dropped hod of clay   = brief spark
                                             = incipient tremor
                                             = separation anxiety
                                             = armistice

          You work in a vacuum,
                    visible
                    as signature wavelengths
                    on a spectrum

We track your movements
        through phases,
        the revolving door
        of question
        and question

August low-pressure system

or

raiment of the superheated,
                non-flammable sort.

4

There from the beginning,
        despite cave bears,
                extinction events.

I sometimes take you for
        that river of forgetfulness,
                your effigy gone to Weehawken . . .

or a fugue state crossing from one palm
        to the next, as deaf mice
                scuttle greyly in the subway bed.

You were here, I am certain of it,
        visible on August 15, 2007, on Google Earth.
                Your wide sunhat. You are jaywalking

and—I would guess—covering your ears
        against construction noise
                or ducking a noisome silence.

You are a shadow parting other shadows
        as though to prove time were the one
                collapsible dimension.

6
C
12.0107

One finger's-worth
constitutes a marsh that stretches

huzzzzzzzz

complete
with cattails and moccasins
and spiny perch
with fluttering gills,
the blood so close
        it might be called upon
                                easily.

O, ready weep.
The drizzle of nomenclature
betrays no sign of letting up.

Fires stand in bones,
urgency
sequestered in the earth's
                        deep seams.

In the long list of nourishments,
        this

           still life
           without
           canary

                    just these
possible selves
in the most unformulated form.

Stipple

Sundown, and the reflection
erupts with hairtrigger mouths,

a frenzy of pumpkinseed,
a Technicolor snow
left on into the evening.

Each angstrom caught—
all scale, tooth and bone—made sod
or buried in the garden plot.

It wells up and runs down
in a hard rain.

Dorsal

A diver follows a pale flash
        through milfoil blooms.

Could be a catfish barb
        or pulled thread of the sainted.

More likely a drawstring, a beer can,
        what's left of the drowned.

Someone keeps the engine running above
        and scans the surface with binoculars.

Ventricle

Rapid,
shallow,

perspiration gathers at the navel
and a high hawk turns.

Trees collapse
in their shade.

Recollection's
faint flutter:

a ragged blue tarp
washes up.

i.

The sound from the culvert
was not nothing.

It might have been the sound
of soldier ants
clenching their mandibles
in their sleep, dreaming

    of swaths of leaf
    for the cutting. Or

stars, perhaps,
skritching across
night's chalkboard.

ii.

In other words,
invocation is not
a one-way street.

iii.

O need, you are afar.

     Still, I would risk
     making your distance near.

iv.

O culvert,
O soldier ant,
O mandible,
O dream.

The stars word you across,
practicing your names.

We started easy, then ratcheted up:
marshmallow chicks to Jonagolds.
Leaf, soil, snail, and later
the contents of a minibar.

Someone got up the notion to try a brick.
Ragged gums, cast-iron constitutions.
Through it all, we smiled:
child bridegrooms on our wedding day.

There was no getting out of it.
We were to be made permanent
when all we wanted was a little blood
and newness working through.

Instead: recurring adolescent dreams
lasting into our twilight. There was
something in the water, alright.

Truth serum elucidates:
the nature of nature
is a white whale. We think
we are the equal to it.

Someone calls out
a name at the party,
and you are instantly convinced
she has perished
in a burning building.

The word 'fate' is a shorthand.
Whatever it is, you have
a knack for it. For finding
the one lure that would
do you in in a whole ocean.

Leading up to disaster,
the captain's log reads:
Once, we were co-conspirators.
Now, the crew is simply
> *wet through, exhausted.*

This was before photography,
which begs the question:

How does one remember
the beloved exactly
as she was, receding on a pier?

Maybe photography
became the crutch,
replacing a perfect
          capacity for the image.

In any event, it is lost.

That name has no more power
on this earth.
We are telling the truth now.
Now.

The room to which you are led
has the aura of seizure, private illumination,
crawlspace behind a cameo or scrim.

    The bitten bench. The lathe he leans
    against. Tall posts for a canopy swing:
    the slashing legs of a thoroughbred.

The last of three sets—which he plans
to give away to his youngest daughter
when she's married off—rests assembled,

    disassembled, stashed
    in the space between ceiling joists
    where it will be discovered

by his wife's next husband, who will take the pieces
for good enough kindling—their splitting
dry deliciousness—and feed them to the stove.

    Here, they'll sing an uncanny *Agnus Dei*
    in sisterly harmonies. The sparks
    winnowing out like midges.

The shore's continual suggestion.
Once a curved green piece arrived,
explained as a Japanese fishing net buoy.

Another ocean. Exactly that. Whitecaps
are furnaces, when you think of it.
The same crisp light, the same drizzling.

One sister gets to know the gone sister's
daughter. Thinking *You remind me of you.*
Light and shade come together in a dovetail.

The green glass thrown back
and a bench at the bottom of the stairs.
Someone else will have to pick the horses

from now on. *Agnus Dei* in green silks.
Or *Dovetail*, perhaps.

for John J.

i.

I awoke this morning with   a frog in my throat
                                  a bell in the frog
                                  a leaf in the bell.

I awoke this morning
     with a bell
     a leaf
     and a frog in my throat,
although not in that order.

So, this is what it feels like
to be there at the start of something.

In on the ground floor
is how I would describe it
to my brother, who works
in the financial sector
     a leaf you could burn
     a bell you could melt
     and a frog you could keep
in the steeple of your hands
or perhaps on your shoulder
like an epaulet—glinting as seen from space—because you goddamned
     thought of it first.

ii.

The last time my brother and I were together,
we got drunk and watched *For All Mankind*.
He was dressed up to play Conspiracy Theorist:
"A roomful of calculators
and they landed on the moon?"
The foil panels were too perfect.
And was that wind ruffling the flag?

But I could only blink
at the veracity of it:
the bell-like module,
the scintillation of Gobi dust.
Urals like the ridges of an open mouth.
Static breaking through Armstrong's voice.

14
Si
28.0855

Early one morning,
the mother of Pliny the Younger
noticed a peculiar
column of cloud
hovering over the strand.

She woke her son
and then the neighbors.

Unsure what to do,
they huddled in the boathouse
while the chained dog
choked slowly on sulfur.

Ghosts full of advice
appeared among the villagers.

The mountain sloughed minerals,
altering the coast.

———

This morning, I felt
left behind,
swaddled in ash, unable
to place the call in the hemlocks.

I had just dreamt a woman
reclining on two lightly gritted elbows
and gazing out at the Atlantic.
Carpel tunnel, vodka spritzers and her shitty job
were the continent behind her.

Your heart
in your ribs
is a fever risking
collapse.

———

Pliny kept reading
while the ashes fell.

The evening beyond each chain-lit match

seemed to crouch in the shapes of houses,
then rose to play havoc in a veil of dogwoods.

In among the lapses, deer stooped
on their stilts to eat the tulips

which, under these circumstances,
turned away from the source

like moths losing themselves in folded wool.

19
K
39.0983

Toward sodium, across a membrane, the molecular equivalent of desire. The living need is Ghost-in-the-Graveyard. I started to describe 'anemone,' but it was gone as soon as I had finished practicing the vowels.

Before you go tearing out of this parking lot, know that the fuses in your glovebox never did anything wrong. Quit laughing. Turn the radio station down. That is not a distant sandbag on sawhorses. It is a fawn wanting—dimly—to cross.

I can't recall the name of the village in western China, only the bronze faces gathered at the basketball court. Boys preening slick hair, girlfriends standing on the seats of their motorbikes for a better look. Bare arms. You expected a fight, not this hush as paper lanterns unfolded and rose on candlepower, uncommonly delicate designs on thin bamboo trusses.

22
Ti
47.867

This is white nested inside a shell of white.

This is how you know the membrane
of the eye is wet, alive.

We are forbidden to watch
the full extent of the eclipse
except through a pinprick shade.

A crest moves across the surface of water,
although the water stays put. It breaks
against enamel between telephonic stutters.

A summons from the folding of a napkin,
our experience of this place through pigments:
rose conniption, silver laughter,
weal of not leaving.
Euphoric bursting of cells,
a marble with its body
opened like a Christ.

Since day six, we have been
digging in with our tongues:

      power tools studded with industrial diamond.

We set off a slick
      under the Yellowstone
and then, for fun, turn stone
      into yellowcake.

You hear that?
         Those are the adoring
    ci-
    CAY-
    das.

Examples of dominion
punctuate our waking hours:

    1) We drive by
        and the field of ethanol corn
        experiences a frisson.

    2) Gregory Peck
        draws a bead
        on a rabid dog.

3) Ecstatic technicians
   carbon date
   the brontosaurus bones.

We are the gods of our own time.
What we can't use, eat or revive,
we cryogenically freeze.

Some of us grow younger in space.
Some of us determine
whether an atom stays
together or falls apart.

Don't talk to us just now.
We are developing a work-around
for the ends of great loves.

27
Co
58.9332

A sponge diver discovered
in 1982
      a blue without precedence
in ingots crossing
from Egypt to Rhodes.

[defiant flapping of sail,
stationary wing in all that waste]

It had gone down
in the Fourteenth Century, BCE,
along with a trove of falcon pendants

      and resurfaced in a suburban garage sale,
      in the guise of a decorative lamp
      displayed on a door between sawhorses
      at the corner of Hemlock and Jones Hollow.
      Ennobled with dust,
      small catastrophe, it stood out
      among the Betamax tapes
      and barbells of absurd increments.

      . . . *Or Best Offer*, thought a packrat,
      turning it in the light.

But it was a child
who finally put down his allowance money
just before the liver-colored clouds
let loose.

And so it was the little soul of blue within
collected him.

28
Ni
58.6934

Rosehipped
the sound after
passes: its own
logic
or stripped,
or reason.

Be the intimate
we tried
on
(Your triad
from a
bas relief.)
solid part
we breathe.

in lowlands.
seemed like the future
open.
it was early
it was crumbling
it was stripped
all jewelweed
bas relief.

subcutaneous
the rain
thirsting
either satisfied
all naked stems

Be reasonable.
landscapes
recreating
a larger scale.
ringing out
crumbling
Be sublimated,
of the air

We walked
Jewelweed
triggering
We sensed
for rosehips
for the reason
for satisfied sounding
rain and triggering

When I have the heatnod,
speckled quince of noonday
sun through closed eyelids,

I think of the Gaul ebbing slowly,
seated on his shield
as though it were a picnic blanket,
and he deciding whether
to begin his meal with the carnyx,
the sword or the leather strap.

A loaf of bread. His bones
already a ribbon of honey
for wolves to lap.

33
As
74.9216

The radio intones that
    A) we have eliminated rinderpest
        and
    B) a number of mushroom hunters are mistaking
        false morels for the genuine thing

My friend says you can tell the difference right away
when your lower half *doesn't* turn
into a bull's nethers.

    Man as bull.
    Bull as man.

But also:
    flower as catastrophe,
    minnow as exception,
    slime mold as jealous light.

I intended to finish the bestiary,
but the cattle of Kenya
kept falling over
with the ache in the side.

And to think
we started off wanting
things to be
just as they appeared.

Pretty soon,
we noticed
the morel has a dusky cousin.

We grew irritable,
the room stifling
with the odor of hay.

No place ~~we had ever~~
~~been~~ together, ~~although~~
familiar ~~wallpaper. One table~~
~~cluttered~~ with antique ~~telephones,~~
sunlight ~~caught up~~
~~in the coiled cords.~~

Ringing ~~throttled~~
~~the house~~ half awake.
~~Understand,~~ it was my ~~job~~
~~to lift each receiver from its~~ cradle
and answer.

[Out of ~~the window:~~ the starling
~~and its discontents,~~
such bankings and syncopated ~~flight].~~

~~"Hello?" I said.~~
~~It was you,~~ calling
after ~~what seemed~~
~~a long absence.~~

~~You said,~~ "It's ~~good~~
~~to hear your voice." Then~~
~~added, nearly as an afterthought,~~
~~that~~ fire ~~had cauterized the field~~

~~where we watched~~ all afternoon
~~the starlings~~ constructing a ~~call
and~~ response.

~~Whose eyes would I believe?~~
And ~~how would I respond
to~~ yet ~~another instance
in which~~ I ~~knew
the universe secondhand?~~

~~Br-r-r-r, Br-r-r-r, it~~ sang.

Gabriel with a weary halo and diminished certainty. Gabriel thinking *I could have been a roofer. Back, pelvis and femur in two places. But on my feet again in time for the dogwoods. In time for the razing of McMansions.* Then, of course, those faulty models—tin fuselage and plastic wings—you wound and wound and let go from the garage. The scientific method was enough for you ... that, and an aerial view of the neighbor's daughter sunning on the porch. Boombox *Club Nouveau* and her father's Budweiser and the cordless like a doll she no longer cared for but made stay, anyway.

38
Sr
87.62

[Above]

The yellow backhoe
leaves the earth open

and a whispering of red
swings out on the hinge.

Remember—my hunger is mortal. I carry sensible provisions:
things that will keep, without seeds. Seeds would have to be spat
somewhere. They stick in the teeth and can be misconstrued for
wanting to stay.

1) strips of mango, dried
2) ampule of a body of water, corked
3) hands like a set of maps

[Below]

anatomizing light of winter

If the nuns of Brest have been here,
I will know them
by the drag of their wool
to and from the houses of the sick

and also the hymns they sing to stay warm.

I will know what I have come for
        when I see it.

a courtyard bounded with stone

Or maybe seeing is the wrong sense.

The scenes in which you feature
have been set afire: Camden Hills in October,
early Hollywood color.

                  The lurid waters
of the bay gain and give ambiguously.
Pinnacles glint: a flash of flesh, a turn of phrase.

Your fear of being alone
is profound in the warehouse district.

Which is not to say the scenes are unpeopled:
consider the family of raccoons
looking on from the storm drain.
Or the children in pioneer dress
texting while guiding hoops
down the leafy neighborhood.

50
Sn
118.710

Standard balloon construction: this frailty
we thought would hold against however many
decibels, whatever category storms,
although we watched cul-de-sacs
bloom overnight, concrete slabs like a river icing up
followed by pilings, joists, studs, a false canopy.

These were the days we dreamt planetary
landings, papier-mâché, solenoids of copper wire coiled
like ringlets in a girlie magazine. Who was it
kept white gas near Southern Comfort?
A rasp and a finer file for sharpening lawnmower blades,
a place for burning the roofs of our mouths,

sunbleaching photographs, secreting extra keys.
A place we could practice the neglect
required to keep on going: droplets rising
to the edges fleeter than sensation,
quartzite pressed into a sleeping hand.

We only ever wanted to gather and cast. Instead,
we were standard balloon construction
leaning into prevailing wind.
It turns out this very afternoon
is a knockdown.

53
I
126.904

a finger of light

                    perched
on a floating raft of ilium

                                        sets it
            gently spinning

lie back
                              you are
            not the sky
                              the raft
                  the river

            the firs
                              are suitors
circling

                                        the sound
            of the heron departing
                  departs

'iodes' from
                  'violet'

            now say
                        what
                  you are

39

56
Ba
137.327

An ache has come to rest
in the harbor of soft tissue

the way the silver of your features—
triangular nose, almond eyes—

came down to you from ancestors
who coughed precisely into handkerchiefs

and stared hard at a hardpan field:
hailstones in the age of the beggar-blind.

Or a cameo at the suprasternal notch.
You want to reassure them

that the stalks were cut and threshed
long before the ceiling of cloud caved in.

> I have your birthmark
> on me here somewhere . . .

> These lips, with their reflexive curl,
> these eye teeth,

a scrap of a verse repeated
like a flag flying in the skull:

l'agneau blanc
et l'agneau noir . . .

She had been
such nightly ragged,
such grown gown
and now—the hour
flown—she felt
her way down

like a ladder of blood,
she felt her way down
while engines purred
in the euphonious early dawn.
Spent selves borne home
through the neighborhood.
She felt her keys—cold—in the pocket
against her thigh.

A toothy muzzle—like a smile—
rested against her thigh.
Her yellowed fingers shook
like a crown-of-thorns.
Why was she now remembering
the useless mnemonic?
She crept up to RECEPTION's window.

Tubes flickered raggedly
above RECEPTION's window.
The toothy muzzle led the way through
cold, and dawn rose like a gown
at the start of a euphonious
purring in the blood.

Had you been born earlier,
you would've had my name.

I remember this
whenever I catch myself
feeling inevitable:
my Dream House,
my Mountains' Majesty.

The truth is, at any moment
my liver could go
to someone more deserving.

My name could apply
to a cellular wish,
tightly curled and hovering
in the pelvic bowl.

We might have been siblings.
Instead, you are backyard salts
and I am standing on my hind legs.
Calling to the denuded valley
like a bear gorged on lichen.

This is why we are acousin,
happy accident of sequencing:
one of us held
the shotgun shell in a small
fist and the other
pried out the pin.

At the dirt market,
the painter of bottles
works small
and from the inside
with brushes
of eyelash.

He takes requests
but also keeps
a stock of cranes,
stoop-necked
and high stepping
behind their screen
of rushes and also

mountain villages
materializing
from the fog.

He draws on a cigarette
as he works closer
to the center.

He has decided never
to paint the girl
fording the stream,

her few possessions
held above her head
as though they were
for the mauve
sky alone.

for Sara

"Slate," he said.

And it *was* late. What happened next was like a barge scudding past a tableau. We turned our heads, and when we looked back our small hairs had been flash-gilded. Insects huzzed in the spikerush. Once dark pines had been ... enpeached.

The problem was current: you existed in two shells, one beside me, and the other far along the surface, so that you were only momentarily visible on the backstroke, upper arms alternating mechanically.

Brief terror, as though plied by wind-ups, governed by terminals and the gates within circuits. Nervy as a finch.

But the powder moon indicated *to be continued*, and something rose to the surface and swallowed as though testing its jaws, not very hungry at all.

A handmade canoe
scuttles in a few feet of water.
The sorriest boat
in the world has gone down.

I sometimes drift over it
on my way to fishing the Cove,
or hover over it, porthole-center
in a truck tire innertube.

On windy days, the float drags
its anchor chains and cinder blocks,
but the sorriest boat in the world
is trim and purposeful, hove to

at the base of Leviathan Rock
with poke grass jutting its ribs,
ready to nose out for deeper green,
which it does for years,

to the astonishment of the mussels,
their flinching purple mouths.

84
Po
208.982

"I would dearly like to bring you back . . ."
—Marie Curie to Irène Curie, August, 1914

Chocolate wrapped in its foil.
Cadences of tinkers in the street.

We could spend an afternoon
calculating for n, and later

training primrose and dropping
maple rotors from the window

onto the city, resplendent
atop its catacombs. Lunch was

half a pomegranate from a stall
in the Place de la Bastille.

Remember when hope was as simple
as division of cells?

———

News reaches me now: a capsule
exploded in your laboratory.

I recognize this land, though
it's estranged and will never

be rid of winter.

These stalled gulls
must imagine
prying the mussels
out by their beards.

Wheeling, clacking
above the bluish clamps,
the orange, orange upholstery
of slumbering fat.

Their angling this afternoon
is acute, only rarer.
Their certain movement,
their certain lack thereof . . .

Once, you were late
getting to the coast
and had to pull to the shoulder.
Exhaustion, relief.

The word for falling
instantly asleep.
You cannot
say it yourself.

Headlands leapt
to a dazzling done.

Gulls disclose the ocean
taking back and back.

| 92 |
| U |
| 238.029 |

Day three of trying to understand Christo's impulse
to line the friable canyon with metallofabric.

We split the atom because we could
and are now outfitting cockroaches with microphones;
our drones have a bird's-eye imagination.

You can still drive past the McMansion on 23
and see where they ran out of loans.
The windows are particle board.
Exasperated or praying for rain.

This was supposed to be the onset of spring;
instead, plain wooden boxes lined with the fabrics
of previous revolutions.

[Remote singing of a lyre
at the vanishing point of a wide vee.]

Green, orange, purple, white.
We did not even have time to name
a flower after this atomic particle.

Much depends upon the disposition of bighorn sheep,
as observed in this bird's-eye photograph.
Notice the individuals shying up
to a languorous stretch of the Arkansas.

The nesting boxes contain
three successive windows
and were fashioned in Colorado Springs.

The cockroaches that survive us will grow
to understand our praying dispositions.

How we were at once so languorous and particular.

We need new art.
And I don't mean a resin toilet.
And I don't mean another naked conversation with the artist.

How long have we been
without alabasters?

The children whose work it is
to take apart CPUs
demand new meaning,
new treatment,
a wall-sized charcoal composition
to make you stumble—Oedipal—
out of the Ashmolean
and into The Eagle and Child, raving
about the last days of empire.

The venom—and someone
will have to help you with this—
the venom comes out best
with a penknife and two little X's.

| 103 |
| Lr |
| 260.105 |

Make a game of sawing through lake ice,
the diatoms aglitter on their maiden voyages.

Your work from here on in
is to think of spells to free them up.

These statements will not be
synonymous with eventual thaw.

A raccoon sleepwalks in broad daylight
through the industrial kitchen.

A silkworm works straight past
its single mulberry meal.

As you cannot throw away this shirt,
roll it up and use it as a bolster pillow.

Tear it to strips for your first aid kit or else . . .
warning flags for following too closely.

I might point out the diatoms
are now as high above as chandeliers

or the staring stars that say the business
of the heart is GOING OUT

57

OF BUSINESS! So, big deal, strain,
blameworthy. But that doesn't help

here in the deepest seams. This is
revival work. Blue of mascara, blue

of five o'clock shadow. Blue of spilt
milk, blue of robin's eggshell.

The look of understanding is plastic,
instantaneous. Someday, Poor

Architect, you will have to leave
behind even the concept of blue,

one purpose being to make
each name strange in the mouth.

105
Db
262.11

[DUBNIUM]

2, 8, 18, 32, 32, 11, 2

subtle

staccato or dithyrambic

bright crests   performative neighbors
and Lifestar helicopter chop
leave us

standing open-mouthed as day lilies
or like mute deer staggering
through the lobby
to forage in the potted palms
now satisfied

and waiting to be shepherded
with flapping hands
or clanking pots
toward the glassed double door
which parts by mysterious mechanism

and closes behind us
on poolshriek   murmurs

resumed

59

Each loosening is without precedent. Resin rises to our mouths and a green gaze gives way, i.e., destruction follows a constructive logic, i.e., our silence is a river to be forded, and the failed attempts—catapult, pendulum, birchbark canoe— slip mutely downstream. Maybe there were no survivors. There were no survivors.

We arose from each other in a scattering of articles—*a, the, an*—definite bodies in the indefinite morning. I was reminded in that new light of the houses, warped and blasted where they stood. The live oak might go on, but its context was gone.

After one is consumed in water, it is difficult to trust solid ground. What we can never be is still: a half-life of remarking the way teeth fit together, starry solutions and naïfs taken in.

We are each remade, wholecloth, at the subatomic level, in the image of an electron.

108

Hs

264

[HASSIUM]

Dependable little asp,
you are as clockwork
as jealousy

       and lead me
like a lurid swathe of color
past bedside vigils
as though dying were
the simplest thing
in our repertoire.

Past the floating cities
on tethers anchored
achingly in our earth.

Past the pink-eyed rabbits
in their sawdust
giving birth and birth.

You are a particle
like the soul—composite, com-
patriot— to walk with as far
as the unlit wood

                    although
I would prefer not to cling
to the old forms: evil seed,
unlucky star.

Wanderer, you are
always home.

118
Uuo
294

By all accounts, we are closer
to a certain grasp

(lash, juncture and midst):

the monarch inherits coordinates;
a subtle loosening makes us tick.

Wait long enough

(mimeograph)

and the carbon comes back to you.

Everywhere one turns:
remarkable likenesses.

You speak to the dresses
in your closet as though
they were children.

The subconscious goes
and makes a deliberate mess
of things; and theory . . .

theory will have to suffice
for now.